INSTANT ICE

Written by
Read With You Science Academy

Read With You

Published by Read With You Publishing
Designed by Read With You Science Academy
Read With You and associated logos are trademarks and/or registered trademarks of Read With You L.L.C.

ISBN-13: 978-1-949391-76-3
ISBN-10: 1-949391-76-0

Second Edition April 2021
Printed in the United States of America.

Welcome to the

READ WITH YOU SCIENCE ACADEMY!

Today we will do an experiment called

INSTANT ICE

Before we begin, let's review some important words:

instantly:
Immediately; right away.

to freeze:
To change something from a liquid to a solid by making it very cold.

to chill:
To make something cold.

SAFETY FIRST

Remember:
Be a careful scientist!
Clean up any water that spills right away!

QUESTION:

What I want to know.

Can a chilled bottle of water freeze instantly when I shake it?

- [ ] Yes, a chilled bottle of water can freeze instantly when I shake it.
- [ ] No, a chilled bottle of water cannot freeze instantly when I shake it.

HYPOTHESIS:

What I think will happen and why.

I think a chilled bottle of water **can** freeze instantly because...

I think a chilled bottle of water **cannot** freeze instantly because...

A chilled bottle of water **can** freeze instantly because:	A chilled bottle of water **cannot** freeze instantly because:
The temperature is low enough.	It takes a long time to make water change to ice.
We can cause a special reaction, or change.	Ice doesn't form by shaking the water.

What do you think? Circle your hypothesis!

EXPERIMENT:

How we will test our hypothesis.

THINGS WE NEED

- 3 plastic bottles of cold water
- Lots of ice cubes
- Salt
- A large bowl
- A glass
- A thermometer
- A watch or timer

bottles of water

ice cubes

salt

bowl

glass

thermometer

watch

PROCEDURE:

The steps to complete the experiment.

Step 1

- Fill the large bowl with ice cubes.

Step 2

Place the three plastic bottles of cold water in the ice. They should be near the center of the bowl, with the ice cubes all around them.
Hint: The ice cubes should go all the way to the top of the bottles!

Step 3

- Put salt across the surface of the ice. You should use a lot!

Step 4

- Place the thermometer into the ice between the bottles.
- For the next 30 minutes, watch the temperature go down, down, down!

Step 5

- The ice will begin to melt, so you will need to add more. Don't forget to add more salt, too!
- The bottles should always be covered in ice, all the way to the top!

Step 6

- Keep watching the thermometer. The temperature needs to go down to 17 degrees Fahrenheit (-8 degrees Celsius).

Step 7

- When the temperature is 17°F (-8°C), set your timer! Keep the bottles in the ice for 10 more minutes, then gently remove one of the bottles from the ice and salt mixture.
- Remember: Clean up any water that spills right away

SAFETY FIRST

Step 8

- Shake, shake, shake the bottle!

Step 9

- Ice may immediately form. Did your water freeze instantly?

Write down what you see

In your science journal,

write what you **observed**!

- What was your **hypothesis**? (Look on page 5 of this book!)
- Was your hypothesis correct?
- Did the ice form in less than 5 seconds?
- Did the ice form at the top of the bottle and quickly move down? Or, did it form at the bottom of the bottle and move up?

Let's do it again!

Did you notice when the ice started forming? Do you think we can make ice form without shaking the bottle? Let's do an experiment to find out. You will need more salt and more ice cubes.

QUESTION:

What I want to know.

Can I make ice form immediately without shaking the bottle?

HYPOTHESIS:

What I think will happen and why.

I **think** I can make ice form immediately without shaking the bottle because…

I **don't think** I can make ice form immediately without shaking the bottle because...

What do you think?

Write your new hypothesis in your science journal.

PROCEDURE:

The steps to complete the experiment.

Step 1B

- Put some ice cubes in the glass.

Step 2B

- Check the thermometer. The temperature should still be 17°F (-8°C).

Step 3B

- Take one bottle of water from the bowl. **Don't shake it!**

Step 4B

- Open the bottle and pour the water on the ice cubes.
- Observe what happens. Did ice form?
- Remember: Clean up any water that spills right away!

Write down what you see

In your science journal, write your results!

- What was your hypothesis? (Look on page 9 of this book!)
- Was your hypothesis correct?
- How did you do the experiment?
- Did the ice form?
- How long did it take the water to turn into ice?

Shake or Pour?	Did ice form?
Bottle #1: Shake	
Bottle #2: Pour	

Why did this work?

- You used salt and ice to lower the temperature of the water below the normal freezing point of water (32°F/0°C). This is called freezing point depression. It means the temperature that water freezes was lowered when you added another compound to it.

- The water in the bottle was supercooled. The temperature of the water was below its normal freezing point, but the water hadn't turned into a solid. That's because the freezing process can't begin until the liquid molecules connect in a regular pattern. They need some help from a rough spot on the bottle or even shaking it.

SCIENCE

- Water needs a **nucleus** so the **crystals** can form and become ice. Usually, water has a lot of particles and **impurities** that can be used to form the nucleus. Bottled water is **purified**, so the water molecules cannot **crystalize** because they do not have anything to use as a nucleus.

- When you shook the bottle, you began the process of **nucleation**. In other words, shaking the bottle created energy, which made a nucleus ice crystal. Then, other ice crystals quickly formed and attached to the first one.

- Slushy ice formed when you poured the supercooled water into a cup of ice cubes. Each ice cube was like a nucleus that helped the water molecules begin the crystallization process. The water doesn't freeze solid because this process releases some heat.

GLOSSARY

Crystals: A small piece of a substance that has many sides and is formed when the substance turns into a solid.

Crystalize: To change into a solid form that is made up of crystals.

Compound: A substance that is formed when the atoms of two or more chemical elements join together.

To freeze: To change something from a liquid to a solid by making it very cold.

Freezing point: The temperature when something freezes.

Freezing point depression: When the normal freezing point of a liquid is made lower by the addition of a solute (substance dissolved into a substance).

Impurities: An unwanted substance that is found in something else and that prevents it from being pure.

Mixture: Something made by combining two or more ingredients.

Nucleation: The process of forming a nucleus.

Nucleus: The central part of an atom that is made up of protons and neutrons.

Purified: To remove dirty or harmful substances from something.

To supercool: To cool a liquid below its normal freezing point without making it a solid.

Surface: The outside part or layer of something.

NEW IDEAS

How could we change this experiment?

- Try taking the 3rd bottle of chilled water and hit it against the table to see if ice will form.
- Take a glass and pour the chilled water in it. Touch the surface of the water with an ice cube. What happens?

Can you think of any ideas to improve the experiment?

- Measure the salt and use the same amount for every test.
- Use the same type and size water bottle.

JOURNAL

Made in United States
Orlando, FL
07 November 2026

90337533R00017